Book 1
Robotics
By Kenneth Fraser

&

Book 2
Human-Computer Interaction
By Solis Tech

&

Book 3
Quality Assurance
By Solis Tech

Book 1
Robotics
By Kenneth Fraser

The Beginner's Guide to Robotic Building, Technology, Mechanics, and Processes!

Robotics: The Beginner's Guide to Robotic Building, Technology, Mechanics, and Processes!

Table of Contents

Introduction

I want to thank you and congratulate you for purchasing the book, *"Robotics: The Beginner's Guide to Robotic Building"*.

This book contains proven steps and strategies on how to build your own robot that will perform certain functions as you want it to do.

For most people, a robot is a machine that could mimic a human such as R2D2 and C3PO in Star Wars. But these types of robots are still in the figments of our imaginations. We are still far from giving robots high level of artificial intelligence to easily adapt and interact to its environment. There is however pioneering works on artificial intelligence that hopes to create humanoid robots.

The type of robots that exist and working today are robots that are programmed to do things that are too dangerous for humans, too repetitive, or just plain messy. These robots are often found in wide range of industries and places such as oil refineries, hospitals, laboratories, factories, and even in the Outer Space. There are about more than a million robots are working in different fields today.

There are types of robots that bring joy to kids such as the popular AIBO ERS-220 that is a bestseller toy during Christmas. While some robots perform great feats by discovering new places and gathering important data in the name of science, specialized robots such as the Mars Rover Sojourner and the underwater robot Caribou are sent to places that average humans cannot go.

Robots are exciting machines to play with, but they are more exciting to build. For hobbyists, building their own robots that capable of doing whatever they program these machines to do gives them pure delight.

This book introduces you to the science of robotics – its basic elements and fundamental concepts. And at the course of your reading, you will learn all the essential aspects you need to build your own robot.

Thanks again for purchasing this book, I hope you enjoy it!

Chapter 1 – What is a Robot?

It is interesting to know that even with all the hype about robots and with all the milestones in robotics, there is still not standard definition for a robot. There are, however, some basic characteristics that a robot should have and this could help you determine if a certain object is a robot or not. It will also guide you in deciding what features you need to build into a machine before you can say that it is a robot.

Four Basic Characteristics of a Robot

A robot has four basic characteristics: sensing, movement, intelligence, and energy.

Sensing

Basically, a robot must be capable of sensing its environment much similar to the way humans sense its surroundings. Robots could either sense through light sensors that mimic the functions of the eyes, or be equipped with chemical sensors that function like the nose, sonar sensors like the nose, touch sensors like the skin, and taste sensors like the tongue. These sensors will help the robot to become aware and understand its environment.

Movement

A robot should have the ability to move around through walking on legs, rolling on wheels, or through propellers. It's either the entire robot is able to move or just some parts of it such as head, arms, or just legs.

Intelligence

A robot should be equipped with artificial intelligence or AI. This is usually done through computer programing. Hence, you need a background in programming to provide your robot with the needed intelligence. You need to program the robot's intelligence so that it will know what to sense and how to move.

Energy

A robot should have a way to power itself. The energy source could be electrical, chemical (battery), or solar. The method by which your robot energizes itself depends on what your robot is required to do.

Working Definition of a Robot

For the purpose of discussion and for reference, we define robot as a machine that contains control systems, sensors, manipulators, software and power supplies that works together to do certain tasks.

Building a robot requires understanding of the fundamental principles of mechanical engineering, mathematics, physics, and computer programming. In special cases, it also requires specific knowledge on chemistry, biology, and medicine. In studying robotics, you need to be actively engaged with wide range of disciplines to build robots that could solve certain problems.

A Brief History of Robotics

The word "robot" was first used in a play entitled R.U.R (Rossum's Universal Robots) written in 1921 by Czech writer Karl Capek. This play is about machines that are built to work on a factory and eventually revolted against their human masters. Robots are the Czech word for slave.

Meanwhile, the word robotics also first appeared in a work of fiction. Russian-born American fictionist Isaac Asimov used it in his short story "Runabout" (1942). Compared to Capek, Asimov had a more positive opinion of the role of robots in the society. In general, he described robots as useful machines that serve humans and perceived them as a "better, cleaner race. He also proposed the three Laws of Robotics:

First Law of Robotics

A robot may not injure a human or, through inaction, allow a human to come to harm.

Second Law of Robotics

A robot must obey the orders given by humans except if such orders would violate the First Law.

Third Law of Robotics

A robot may protect itself as long as such protection do not violate with the First Law or Second Law of Robotics.

Early Models of Robots

Among the earliest cases of a mechanical system designed to perform a regular task was recorded around 3000 BCE. Egyptian water clocks are added with human statuettes to hit the hour bells and signal the passing of time. In 400 BCE,

Archytus of Taremtum, who was known as the inventor of pulley and screw, created a pigeon made of wood that is capable of flying. Meanwhile, hydraulically-powered figurines that could speak prophecies were common during the Greek domination of Egypt during the second century BCE.

In the first century C.E., Petronius Arbiter built a doll that is capable of moving like a human being. In 1557, Giovanni Torriani built a wooden robot, which could fetch the Emperor's bread every day from the store. By 1700s, robotic inventions became common with numerous impractical yet ingenious machines such as steam-powered automata crafted in Canada as well as the popular talking doll by Thomas Edison. Even though these creations may have inspired the design and functions for the modern robot, the progress during the 20th century in the field of robotics exceeded previous advancements many times over.

The First Modern Robots

The robots that we are familiar with were built by George C. Devol in the 1950s. The inventor from Kentucky designed and patented a reprogrammable manipulator that he dubbed as "Unimate" derived from "Universal Automation." For years, he tried commercializing his product, but failed. But in 1960s, the entrepreneur-engineer Joseph Englberger bought the patent from Devol and modified it into an industrial robot. He established his company, Unimation, for production and marketing of these products. He was successful in this venture, and in fact, Englberger is regarded today as the Father of Robotics.

Robotics also progressed within the academic institutions. Alan Turing, pioneering computer scientist, mathematician, logician, and cryptologist, published his book "Computing Machiner and Intelligence" where he proposed a test to determine if a machine has the capacity to think for itself. This test is known as the Turing Test.

In 1958, Charles Rosen of the Stanford Research Institute created a research team to work in the development or a robot known as "Shakey" that was more advanced compared to Devol's Unimate. Shakey can move around through the room, sense light through his "eyes" move around strange environment, and to a particular degree, and react to what is happening to his surroundings. He was called Shakey because of his clattering and rickety motions.

In 1966 at Massachusetts Institute of Technology (MIT), Joseph Weizenbaum created an artificial program named ELIZA, which functions as a computer psychologist that manipulates its user's statements to formulate questions.

In 1967, Richard Greenblatt developed MacHack, a program that is capable of playing chess, as a response to a critical article written by Hubert Dreyfuss where he boasted that no computer program can beat him in chess. When the program is finished, Dreyfuss was invited to play and was defeated. This program was the foundation of future chess programs that eventually developed into Big Blue, the program that defeated Grand Master Gary Kasparov in 1997.

The interest in robotics is one of the major catalysts in the development of computers. In 1964, the IBM 360 becomes the first computer to be produced massively.

Robots are also crucial in pioneering space explorations. In 1969, the United States successfully used the latest technology in robotics and computing for Neil Armstrong's landing on the moon. Robots also helped in the expansion of scientific knowledge. In 1994, Carnegie Universities crafted Dante II, an eight-legged walking robot that successfully descends into the crater of Mt Spur to gather samples of volcanic gas.

Commercial companies also leveraged on the mass appeal of robots. In 1999, Sony released its original version of AIBO, a robotic dog that can entertain, learn, and communicate with its owner. Advanced versions have followed in the succeeding years, with the final model, the ERS-7M3, released in 2005.

Honda also released its ASIMO robot, an advanced humanoid robot in 2000. In 2004, Epsom was hailed as the world's smallest robot (7 cm high and weighs only 10 g.) The robotic helicopter is designed to fly and capture videos during natural disasters.

After being released in 2002, a robotic vacuum cleaner known as the Roomba became a huge hit. It sold more than 2.5 million units, which shows that there's really a huge demand for domestic robot technology.

Hundreds of films feature robots such as The Day the Earth Stood Still (1951), Arthur C. Clark's 2001: A Space Odyssey (1968), Star Wars (1977), Blade Runner (1982), Terminator (1984), Nemesis (1992), I, Robot (2004), Transformers (2007), and many more. The popularity of robotic films shows that people are inspired and delighted by the idea of machines that can independently move and think for itself.

If you are ready to build your own robot, continue to the next Chapter to help you get started.

Chapter 2 – Get Started

The first step in building your own robot is to determine what it should do, that is, your purpose of why you are building the robot. Robots can be used in different situations and are mainly designed to assist humans. It will help you a lot to learn first the different purposes and uses of robots.

Basically, robots are divided into two main groups: industrial and domestic robots.

Industrial Robots

Industrial robots are used in factories to manufacture products with precisions such as computers, cars, cellphones, medicine, and even food. Robots increased the productivity in different workplaces, which resulted to booming industries. Each type of industrial robot has its specific form that corresponds to its function. For instance, robotic arms are often used in car assembly lines to spray paint or weld frames. Robotic arms are among the most common robots today. Recently, agricultural robots have been introduced mainly to perform farm tasks such as cutting weeds and harvesting crops.

Domestic Robots

Domestic robots are mainly used in the home to perform household chores. They usually perform repetitive tasks every day such as vacuuming floors, mowing the lawn, vacuuming floors, and other chores that people usually don't have time to do. For example, there are vacuum robots that can clean the floors. They are equipped with motion sensors so they will not run into any object. You just need to push the switch on and it will do its job. It could pick up dust and pet hairs and could be used for hours.

There are also mower robots that could mow lawns. They are equipped with sensors to detect grass edges. Domestic robots are also used for entertainment such as Robosapien, AIBO, and iDog.

Choosing a Robotic Platform

The next step in building your own robot is to decide on the type of robot you want to build. A usual robot design usually begins with "inspiration" of what the robot will do and what it will look like.

The types of robots that you can build are endless. As long as you can envision something that a robot can do, you can work your way to achieve it. But for beginners, you can start with the following types: land robots, aerial robots, aquatic robots, stationary robots, and hybrid robots.

Robotics: The Beginner's Guide to Robotic Building, Technology, Mechanics, and Processes!

Land Robots

Land-based robots, particularly those added with wheels are among the most common mobile robots built by beginners, because they often require minimal investment while providing the opportunity to learn more about robotics. Meanwhile, the most advanced type of robot is the humanoid robot, which is akin to humans. Humanoids require several degrees of freedom and synchronization of different motors and use several sensors.

Wheeled Robots

Wheels are among the most common method of adding mobility to a robot and are used to mobilize many different sizes of robots and robotic platforms. Wheels could be about any size, and there's no limit in the number of wheels that you can add. More often than not, robots that are equipped with three wheels are using two wheels and a caster at one end. More advanced robots with two wheels are using gyroscopic stabilizing technology.

Meanwhile, robots that are added with four to six wheels usually use several drive motors that decreases the risk of slippage. Also, mecanum wheels or omni-directional wheels can provide the robot considerable benefits in mobility. Most beginners in robotic building are mistaken in thinking that inexpensive DC motors can mobilize robots that are medium in size. As you will learn later, there are more factors that you need to consider before you can add mobility to your robot.

Advantages

Wheeled robots are ideal for beginners as they are often more affordable to build. They have simple design and construction, and there are unlimited options. In addition, robots with six wheels or more could rival the mobility of a track system.

Disadvantages

Wheeled robots usually have small contact area, because only a small portion of the wheel is touching the ground. This results to lower traction that may cause slippage.

Tracked Robots

Tracks are used in tanks for mobility. Even though tracks, also known as treads, don't provide the added torque, they can decrease slippage and can equally distribute the robot's weight. This makes the robot easier to mobilize in loose ground such as gravel and sand. In addition, flexible track systems could easily navigate through a bumpy surface. Most hobbyists also believe that tank tracks are quite cool compared to wheels.

Advantages

Steady contact with the ground avoids slippage, which is prevalent with wheels. The track system also distributes weight evenly, which helps the robot in navigating different surfaces. Tracks can also be used to extensively enhance the ground clearance of the robot without adding a bigger drive wheel.

Disadvantages

The main disadvantage of using a track system for robots is that in turning, there's the tendency to cause damage to the surface that also causes damage to the tracks. In addition, robots are often built around the tracks, and there's a limit in the availability of the tracks. Drive sprocket can also considerably restrict the number of motors that you can use.

Legged Robots

More and more robots are using legs for movement. Legs are usually ideal to use for robots that should navigate on uneven ground. Many prototype robots are built with six legs that allow the robot for static balance. Robots with fewer legs are more difficult to balance as it requires dynamic stability. Once the robot ceases moving in the middle of the stride, it could fall over. Even though there were robots with one leg moving by hopping, bipeds, quadrupeds, and hexapods are the most common forms.

Advantages

The leg motion is the most natural among the platforms, and it can easily overcome big obstacles and move through rough surface.

Disadvantages

Most beginners are discouraged in building their first robot that moves using legs, as it requires high level of electronic, mechanical and coding skills. You also need to find a small battery that can provide the required power, so legged robots are usually expensive to build.

Aerial Robots

Humans have long been inspired by the idea of flight, and this transcends into the field of robotics. The idea of Autonomous Unmanned Aerial Vehicle (AUAV) has gained popularity over the years, and many enthusiasts have developed numerous prototypes. However, the benefits of crafting aerial robots have yet to prevail over the disadvantages. In building aerial robots, many hobbyists are still using commercial remote controllers. Professional aircrafts such as the Predator commissioned by the US military were partially autonomous though recently, updated versions of the Predator have completed aerial missions with only minimum human intervention.

Advantages

Aerial robots are great for surveillance, and remote controlled aircraft has been developed through the years, so there is a diverse community for mechanics where you can find support and know-how in building your own aerial robots.

Disadvantages

There is still limited community when it comes to autonomous control, as most of the knowledge on this field is protected by the US military. Meanwhile, this robot type is expensive as the whole robot could be broken if you miscalculate the steps and lead the robot into a crash.

Aquatic Robots

Recently, more and more hobbyists, communities, and companies are building unnamed aquatic vehicles. There are still many hindrances to overcome in order to make aquatic robots more enticing for the wider communities in robotics. But it is interesting to take note that there are companies today who are manufacturing robots that can clean pools. Aquatic robots can use thrusters, ballast, wings, tails, and fins to move under water.

Advantages

A massive part of the ocean is still unexplored so there's a lot to discover if you choose to build aquatic robots that could help in discovering the underwater world. The robot design is also guaranteed to be unique, and it could be tested in a pool.

Disadvantages

Aquatic robots are often very expensive to build, and there is the risk that the robot could be lost while deep in the ocean. You should also take note that most electronic parts don't pair well with water, especially salty water. You also need to consider the water pressure as going beyond deep sea needs considerable investment and research. There is also very limited robotic community that can provide support, and also limited wireless communication options.

Hybrid Robots

Your concept for the robot may not easily fall into any of the categories mentioned above or could be composed of various functional components. Take note that this book is written to guide you in building mobile robots and not those with fixed designs. In building a hybrid design, it is best to use a modular design where each functional component could be taken off and tested as a separate part.

Advantages

Hybrid robots are designed and built according to your preferences and needs. These robots could be used for various tasks and can be composed of modules. Hybrid robots could lead to versatility and increased functionality.

Disadvantages

Hybrid robots are often complicated to build and expensive. Parts need to be customized to fit the design.

Grippers and Arms

Even though grippers and arms don't fall under the category of mobile robots, robotics basically began with end-effectors and arms. Grippers and arms are the most ideal way for a robot to interact with the environment it is dealing with. Basic robotic arms could have just two to three motions; while more advanced arms could have more than a dozen movements.

Advantages

Most robotic arms and grippers have simple designs, and it is easy to make a three to four degree of freedom robotic arm with a turning base and two joints.

Disadvantages

Robotic arms are stationary unless you fix them on a mobile platform. The cost of building arms or grippers depends on the lifting capacity you need.

In the next chapter, you will learn how to choose the right actuators or motors for your robot.

Chapter 3 – Understanding Actuators

After learning general information about robots and robotics in the first two chapters, it is now time to choose the right actuators to mobilize your robot.

What Are Actuators?

Actuators are devices that transform energy into physical motion. In robotics, this energy is usually electrical energy. Most actuators today produce either linear or rotational motion. For example, a DC motor is a type of actuator.

Selecting the right actuator for your robot requires learning the available actuators, and some fundamental knowledge of physics and mathematics.

Rotational Actuators

Rotational actuators convert electrical energy into rotating motion. There are two primary mechanical parameters that distinguish each actuator: (a) the rotational speed that is often measured in revolutions per minute or rpm and (b) torque or the force that the devices can produce at a given distance often expressed in Oz•in or N•m.

AC Motor

Alternating Current (AC) is rarely used in robots because most of them are powered through Direct Current (DC) in form of cells or batteries. In addition, electronic parts use DC, so it is easier to use the same type of power supply for the actuators. AC motors are primary used in industrial settings where high torque is necessary or where the motors are connected to a wall outlet.

DC Motor

DC motors are often cylindrical in shape but they also come in different shapes and sizes. They also have output shafts that rotate at high speed often between 5000 and 10000 rpm. Even though DC motors rotate very fast, most have low torque. To decrease the speed and add torque, a gear could be added. To install a motor into a robot, you must fix the body of the motor to the robot's frame. Hence, motors usually have mounting holes that are basically located on the motor's face so that they can be easily installed. DC motors could either rotate in counter clockwise or clockwise. The angular movement of the turning shaft could be measured using potentiometers and encoders.

Geared DC Motor

A DC Motor could be added with a gearbox to reduce the motor's speed and enhance its torque. For instance, if a DC motor rotates at 5000 rpm and produces a 0.0005 N•m of torque, adding a 123:1 ("one hundred and twenty three to one") gear would reduce the speed by a factor of 123 (resulting to 5000 rpm / 123 = 40 rpm) and increase the torque by a factor of 123 (0.0005 x 123 = 0.0615 N•m). The most common types of gears are planetary, spur, and worm. Similar to a DC motor, a geared DC motor can also rotate in either clockwise or counter clockwise. You can add an encoder to the shaft if you want to know the number of rotations of the motor.

Hobby Servo Motors

Hobby Servo Motors, also known as R/C Servo Motors are actuators that rotate to a certain angular position, and were traditionally used in more expensive remote controlled machines for controlling or steering flight surfaces. Today, they are used in different applications so their prices have been reduced considerably, and the variety has also increased. Most servo motors can only rotate about 180 degrees. A hobby servo motor is composed of a DC motor, electronics, gears, and a potentiometer that measures the angle. The latter works with the electronics to mobilize the motor and stop the output shaft at a certain angle. In general, these servos have three wires, voltage in, control pulse, and ground. A robot servo is a recently developed servo that provides both position feedback and continuous rotation. Servos could rotate clockwise or counterclockwise.

Stepper Motors

As the name implies, stepper motors rotates following certain steps or degrees. The number of degrees the shaft rotates with every step could vary depending on various factors. Majority of stepper motors don't include gears, so similar to a DC motor, the torque is quote low. Fixing gears to a stepper motor has similar effect as installing gears to a DC motor.

Linear Actuators

Linear actuators produce linear movements. They have three primary distinctive mechanical properties: (a) the force measured in kg or lbs (b) speed measured in m/s or inch/s and (c) the maximum and minimum distance that the rod could move also known as the stroke measured in inches or mm.

Linear DC Actuator

A linear DC actuator is usually composed of a DC motor attached to a lead screw, which also turns as the motor moves. The lead screw has a traveler that is forced either away or towards the motor, basically transforming the rotating motion to a linear movement. Some DC linear actuators integrate a linear potentiometer that adds a linear position feedback.

Solenoids

16

Solenoids are comprised of a coil wound surrounding the mobile core. Once the coil is energized, the core is forced away from the magnetic field and creates a motion in one direction. Several coils or some mechanical arrangements will be needed to provide movements in different directions. A solenoid stroke is often very small but they are often very fast. The strength primarily depends on the size of the coil and the electrical power passing through it.

Hydraulic and Pneumatic Actuators

Hydraulic and pneumatic actuators use liquid or air respectively to create a linear movement. These actuators could have lengthy strokes, high speed and high force. To use these actuators, you need to use a fluid or air compressor that makes them harder to use compared to basic electrical actuators. These are often used in industrial applications because of their large size and high force speed.

Muscle Wire

Muscle wire is a specialized wire, which contracts when electricity passes through it. When electricity is gone and once the wire cools down, it will go back to its original length. This type of actuator is not fast, strong, or creates a long stroke. Nonetheless, it is one of the most convenient actuators to use if you need to work with smaller parts.

How to Choose the Proper Actuator for Your Robot

To guide you in choosing the actuator for certain tasks, consider answering the following questions to help you.

Take note that new innovations and technologies are always being released regularly, so nothing is permanent. Also remember that one actuator could perform various tasks in various contexts.

1. Do you need to mobilize a wheeled robot?

Drive motors should carry the weight of the whole robot and will most likely need a gear down. Majority of the robots utilize "skid steering" while trucks or cars utilize rack and pinion steering. If you prefer the skid steering, geared DC motors are recommended to use for robots with tracks or wheels. Geared motors provide constant rotation, and could have discretionary position feedback through optical encoders. Because the rotation needed is limited to a certain angle, you can choose a hobby servo motor for stirring.

2. Is there a limit on the range of motion?

If the range is restricted to 180 degrees and the needed torque is not a critical factor, a hobby servo motor is recommended. Servo motors are available in various torques and sizes and comes with angular position feedback. Majority of these motors use potentiometer, while some specialized ones use optical encoders. R/C servos are now popularly used to build small walking robots.

3. Do you need a motor to lift or turn heavy loads?

Raising a weight needs considerably more power compare to moving a weight on a flat surface. Torque should be prioritized than the speed, and it is ideal to use a gearbox with a powerful DC motor or a linear DC actuator with a high gear ratio. You can use an actuator system that could prevent the mass from falling if there is a disruption in the power source. This includes clamps or worm gears.

4. Do you need the angle to be precise?

Stepper motors that are paired with a motor controller cold provide a very precise angular motion. They are more ideal to use compared to servo motors because they provide constant rotation. But there are also high-end digital servo motors that use optical encoders and can provide high precision.

5. Do you need to achieve movements in a straight line?

Linear actuators are ideal for moving parts and placing them in a straight line. They are available in different configurations and sizes. For fast movements, you must consider solenoids or pneumatics, for high torques, you can use linear DC actuators or hydraulics, and if the movement requires minimum torque, you can use muscle wire.

Chapter 4 - Microcontrollers and Motor Controllers

Microcontrollers are considered as the "brain" of the robot because it is responsible for all decision making, computations, and communications. These are devices with the capacity to execute a program (a series of instructions).

To interact with the external world, a microcontroller has a sequence of electrical signal connections (known as pins), which could be switched on or off using programming functions. These pins are also used in reading electronic signals that are released by sensors or other devices and determine if they are low or high.

Majority of microcontrollers today could measure analogue voltage signals, or signals that could have a full range of values rather than just two specified states by using analog to digital converter or ADC. Through the use of ADC, a microcontroller could assign a numerical value to the analog voltage that is neither low nor high.

What Could Microcontrollers Do?

Numerous complicated actions could be achieved by setting the pins low and high creatively. Nonetheless, building complicated algorithms such as smart movements and data processing or complicated programs are not yet on the range of microcontrollers because of its natural speed and resource limitations.

For example, to light a blinker, you can program a repeating sequence in which the microcontrollers could turn a pin high, wait for several seconds, turn it low, wait for several seconds and goes back to the first sequence. A light that is connected to the pin will then blink open-endedly.

Similarly, microcontrollers could be used to take control of other electronic devices including actuators when they are installed to motor controllers, Bluetooth or WiFi interfaces, storage devices, and many more. Because of its versatility, microcontrollers could be found in common everyday products. Basically, every home electrical device or home appliance utilizes at least one microcontroller.

Not similar to microprocessors found in Central Processing Units in personal computers, microcontrollers don't need peripherals such as external storage devices or external RAM to operate. Hence, even if the microcontrollers are less powerful compared to microprocessors, building circuits and products based on microcontrollers is an easier task and a lot more affordable, because minimal hardware parts are needed. Remember, microcontrollers can output minimum amount of electrical power through pins. Hence, a generic microcontroller cannot power solenoids, power electrical motors, large lights, or other direct loads. Doing this could cause physical damage to the controller.

Programming Microcontrollers

There's no need to shy away from programming microcontrollers. Not similar in the past where making a blinker took comprehensive knowledge of microcontroller and at least a dozen line of code, programming microcontrollers is fairly easy today. You can use the simplified Integrated Development Environments (IDE), which uses modern languages, full line archives that could cover all of the most common actions, and several handy samples to help you get started. You can learn more about programming your robot in Chapter 6.

How to Choose the Proper Microcontroller for Your Robot

You will need a microcontroller for any robotic building project unless you're into BEAM robotics or you want to control your robot through an R/C system or a tether. For starters, selecting the right microcontroller could seem like a difficult job, particularly considering the product range, specifications, and applications. There are various microcontrollers available today such as BasicATOM, POB Technology, Pololu, Arduino, BasicX, and Parallax.

The following questions could guide you in choosing the right microcontroller:

1. Which microcontroller is widely used for your type of robotic project?

Building robots is not a popularity contest, but the fact that a microcontroller has a large supporting community or has been used in the same project can make the design phase easily. With this, you can benefit from other experience and hobbyists. It is common for hobbyists to share codes, pictures, instructions, and videos even lessons learned.

2. Do you need specific accessories for a certain microcontroller?

If your robot has special needs or there is a certain accessory or component that is important for your design, selecting a compatible microcontroller is clearly essential. Even though most accessories and sensors could be directly interfaced with most microcontrollers, some accessories are designed to interface with a particular microcontroller.

3. Do you need special features for your robot?

A microcontroller should be able to perform all the special actions needed for your robot to function well. Some features are common to all microcontrollers such as being able to execute basic mathematical operations, having digital inputs and outputs, and making decisions. Others may need certain hardware such as PWM, ADC, and communication protocol support. You must also consider pin counts, memory and speed requirements

Motor Controllers

Motor controllers are electronic devices that serve an intermediary device between a microcontroller, the motors, and the power supply.

Even though the microcontroller decides the direction and the speed of the motors, it doesn't have enough power to directly drive them. Meanwhile, the motor controller can supply the current at the needed voltage but doesn't have the capacity to decide how fast the motor must turn.

Hence, the microcontroller and the motor controller must work together to make the motors move accordingly. The microcontroller can provide instructions to the motor controller on how to power up the motors through a standard and basic communication method such as PWM and UART. In addition, some motor controllers could be manually regulated using an analog voltage often created through a potentiometer.

The size and weight of a motor controller may greatly vary from a device that is smaller than the tip of a pencil to a huge controller that could weight several kilos. The size and weight often has a minimum effect on the robot, unless you want to build unnamed aerial or aquatic robots.

Types of Motor Controllers

Because there are several types of actuators (as we have discussed in Chapter 3), there are also several types of motor controllers: brushed DC motor controllers, brushless DC motor controllers, servo motor controllers, and stepper motor controllers.

How to Choose a Motor Controller

You can only choose a motor controller after you have decided on what type of actuator you want to use. In addition, the current that a motor draws depends on the torque it could provide. A small DC motor will not use much current, but cannot also release much torque, while a bigger motor could release higher torque but will need increased current.

Chapter 5 - Controlling Your Robot and Use of Sensors

Based on our definition of a robot, it should gather data about its surroundings, make smart decisions and then execute actions based on calculations. This also includes the option for the robot to become semi-autonomous (with aspect that are controlled by humans and other aspects that it can do on its own).

One good example of this is a complex aquatic robot. A human controls the basic motions of the robot while an installed processor measures and reacts to the underwater currents to keep the robot in one position while still preventing a drift. A camera installed in the robot would send videos back to the human while the sensors could track the water pressure, temperature, and more. Once the communication line falters between the robot and the human, an autonomous program could take over to instruct the robot to reach for the surface.

In controlling your robot, you need to figure out the level of autonomy. You need to choose if you want the robot to be tethered, wireless, or autonomous.

Tethered

Direct Wired Control

The simplest way to control a vehicle is by using a handheld controller that is physically connected to a vehicle using a tether or a cable. Knobs, switches, joysticks, buttons, and levers on the controller will allow you to control the robot without the need to add sophisticated electronics. In this setting, the power source and motors could be directly connected with a switch to control the rotation. These machines often have no artificial intelligence and are regarded as remote controlled devices than robots.

Wired Computer Control

Another method is to integrate a microcontroller into the machine but still using a tether. Attaching the microcontroller to your computer's ports will allow you to control the actions using the keyboard, a joystick, a keypad, or other device. Adding a microcontroller to your robot project may also require programming how the robot will respond to the input.

Ethernet

Another way to use computer control is to use an Ethernet interface. A robot that is directly connected to a router can also be used for mobile robots. Building a robot, which can communicate through the internet could be sophisticated, and usually a wireless internet connection is more recommended.

Wireless

Infrared

You can ditch away cables and wires if you use infrared transmitters and receivers. This is often a great achievement for beginners. Infrared control needs "line of sight" to function. The receiver should have the ability to see the transmitter to receive the data. Infrared remote controls can be used to send commands to infrared receivers that are paired with microcontrollers that interpret these signals and control the actions of the robot.

Radio Frequency

Remote control units often use microcontrollers in the receiver and transmitter for data transmission through radio frequency. The receiver box usually has a printed circuit board (PCB) that includes a small servo motor controller and a receiving unit. RF communication needs a transmitter matched with a receiver or a transceiver. RF doesn't need clear line of sight and could also provide considerable distance. Basic RF devices could allow for data transfer between devices between long distances, and there's no limit to the range for more RF devices.

Bluetooth

Bluetooth is a type of Radio Frequency and follows certain protocols for sending and receiving data. Standard Bluetooth range is usually restricted to about 10 meters although it has the advantage of controlling the robot though Bluetooth-enabled devices including laptops, smartphones, and PDAs. Similar to RF, Bluetooth provides two-way communication.

WiFi

Recent development in wireless technology enables you to control a robot through the Internet. To build a WiFi robot, you must have a wireless router that is connected to the internet and a WiFi device on the robot. You can also use a device, which is enabled with TCP/IP with a wireless router.

Autonomous

High-level robots are autonomous. With recent developments, you can now use the microcontroller in its full potential and program it to respond to input from the sensors. Autonomous control may come in different types: restricted sensor feedback, pre-programmed with no feedback from the environment, and complex sensor feedback. Genuine "autonomous" control includes different sensors and code to allow the robot to figure out by itself the smartest action to be taken in any situation.

The most sophisticated methods of control presently used on autonomous robots are auditory and visual commands. For auditory control, a robot will react to the sound of the human's voice for instructions such as "get the ball" or "turn left." For visual command, a robot may look to an object to decide on what to do. Instructing a robot to turn to the right by showing a drawing of an arrow that is pointing to the right requires complicated programming. Even though these things are no longer impossible, they need a sophisticated level programming and usually hundreds of hours.

Not similar to humans, robots are not restricted to just sound, sight, smell, touch, and taste. Robots use different electromechanical sensors to understand and discover their surroundings. Mimicking a natural organism's senses is presently a great challenge, so developers and robotic builders are using alternatives to these natural senses.

Chapter 6 - Assembling and Programming a Robot

After learning all about the fundamental blocks in building a robot, the next stage is the designing and building of the frame that will keep all the components together and will provide your robot a definite look and shape.

Constructing the Frame

There's no fix method in creating a frame, because there is often a trade-off to be constructed. You may prefer to use a lightweight frame but you may need to use costly materials. You may like a strong or big chassis but you may realize it is expensive, hard, or heavy to produce. The frame could be complicated and may take some time to design and build.

Materials

There are different materials that you can use in creating a frame for your robot. As you try different materials to construct not only robots but other types of machines, you will also understand the advantages and the disadvantages as to which material is the most suitable for a specific project. The roster of suggested building materials below comprises only the more common one, and when you have tried several of these materials, you can start experimenting or blending some together.

Basic Construction Materials

Some of the most basic construction materials could be used to build good-quality frames. The cheapest materials is the cardboard that you can usually find for free and could easily be bent, cut, layered, or bent. You can construct a reinforced cardboard box that looks a lot nicer and is more proportional when it comes to the size of your robot. You can then paint it with glue or epoxy to make it stronger then add extra layer of paint.

Structural Flat Materials

For a more durable frame, you can use a standard structural material such as a sheet of plastic, metal, or wood. You just need to puncture some holes to connect the electronic components. A stronger piece of wood has a tendency to be heavy and thick, while a thin sheet of metal could be too flexible. You can attach components to both sides and the wood will still remain solid and intact.

If you're at the stage where you are ready to have an outsourced frame, the best option is to acquire the part precision cut through a water or laser jet. Hiring a third-party to produce a custom part is recommended only if you are completely sure of the dimensions, because the mistakes could be expensive. Companies that offer computer controlled cutting services may also provide different other services such as painting and bending.

3D Printing

Building a frame constructed from 3D printed panels is not always the most structurally sound option, primarily because it is built up in several layers. However, this process could produce complex and detailed shapes that could be impossible to build using other methods. One 3D printed component may contain all the important mounting points for all mechanical or electrical parts without compromising the robot's weight. In the past decade, the cost of 3D printing is quite expensive, but as it becomes popular, the price of producing the components is also expected to go down.

Assembling the Parts Together

With the available options for materials and methods, you can now start assembling the parts together. You can follow the steps below to build a simple, aesthetic, and structurally reliable robot frame.

1. Decide on the material you want to use.

2. Gather all the parts that your robot will need, both mechanical and electrical and measure them. In case you don't have all the components ready, you can refer to the dimensions that are often supplied by the manufacturers.

3. Think of and draw various designs for your frame. It's fine not to provide details.

4. Once you find the suitable design, be certain that the structure is reliable and all the parts would be supported in the frame.

5. Sketch every component of the robot on cardboard or paper at true scale. You can also draw the parts in the CAD software and print them.

6. Test the design in CAD and in actual setting using your paper prototype by test fitting every component and connection.

7. Measure the dimensions again and when you are completely certain that your design is right, begin cutting the frame into the material. Take note to measure two times and only cut once.

8. Test fit every part before assembling the frame if in case you need some changes.

9. Construct the frame using appropriate assembling materials such as glue, nails, screw, duct tape, or any appropriate binding tools that you prefer.

10. Fit all the parts into the frame and there you go, you have just built your robot!

Constructing the Robot Parts

The last step discussed above should be described further. In the past chapters, you have already chosen the electrical parts including the actuators, microcontroller, and motor controller. The next step is to construct them so they will work together.

In the following section, we'll use standard cable colors and terminal names that encompass common parts. You must rely on manuals and datasheets when you are working on your specific parts.

Attaching Motor Controllers to Motors

A geared DC motor or a linear DC actuator usually has two wires: black and red. Attach the black wire to the M- terminal on the DC motor controller and the red wire to the M+ terminal. Connecting the wires the other way around will only cause the motor to rotate in the opposite direction. Meanwhile, servo motors have three wires: red, black, and yellow. A servo motor controller comes with pins that are matching these wires so you can just plug it directly.

Attaching Microcontroller to Motor Controllers

Microcontrollers can communicate with motor controller in different ways: 12C, R/C, Serial, or PWM. Be sure to refer to the manual for each microcontroller for specific instructions on proper connection. Regardless of the method you choose, the microcontroller and the logic of the motor controller should share matching ground reference. This can be achieved by attaching the GND pins together. Meanwhile, a logic level shifter is needed if the devices don't share the same logic levels.

Attaching Batteries to a Microcontroller or a Motor Controller

Majority of the motor controllers available today have two screw terminals for the battery labels marked with B- and B+. If the batteries you got are provided with a connector and the controller comes with screw terminals, you could still search for a pairing connector with wires that you can attach to the screw terminal. If this is possible, you need to find another way to link the battery to the motor controller while you can still unplug the battery and link it to a charger. It's possible that not all the electrical and mechanical components you have selected for your robot could operate on a single voltage, and so may need several voltage regulation circuits or batteries.

If you are building a robot with a microcontroller, DC gear motors, and maybe servo motors, it's easy to see how a battery may not be able to power every component directly. Nevertheless, it is best to choose a battery that can directly power as many devices that you need. The battery with the largest capacity must be connected with the drive motors. For instance, if the motors you select are rated a nominal 12 volts, the primary battery must also be 12 volts. So you can use a regulator to energize a 5 volts microcontroller. LiPo and NiMH batteries are the top choices for small to medium robots. Select NiMH if you need cheaper batteries and LiPo if you need light weight batteries. Always take note that batteries are powerful devices that could easily burn your circuits if they are not properly connected. Always make sure that the polarity is correct and that your device could handle all the energy supplied by the battery. If you are not certain, never make assumptions.

Adding Electrical Parts to Frame

You can attach electrical components to your frame through different methods. Make certain that whatever method you use, don't conduct electricity. Usual methods include screws, hex spacers, Velcro, double-sided tape, cable ties, glue, and many more.

Programming Your Robot

Programming is often the last step in building your robot. If you have followed the steps described in the previous chapters, by now you have selected the electrical components such as actuators, microcontrollers, motor controllers, sensors, and more. At this point, you might have already constructed your robots and hopefully it looks something like you want it to be. But without the proper program, your robot is just a cool paperweight.

It requires another book to teach you robotic programming. Instead, this section will guide you on how to get started and what you should learn.

There are several programming languages that you can use to program the microcontroller that will serve as the brain of your robot. The following are the most common programming languages you can choose:

Assembly

This programming language is just a shy away from programming a full-pledged computer, and so it could be difficult to use. This language is ideal to use if you really need to ensure complete instruction-level control of your robot.

Basic

Basic is one of the most common programming languages for robot hobbyists. This is often used in programming microcontrollers mainly for educational robots.

C++

C++ is a very popular programming language. It provides top=level functionality while you are keeping a good low-level control. A variant of C++ is Processing, which includes simplified codes to make the programming easier.

Java

Java is more developed compared to C++ and offers any safety features to the disadvantage of low-level control. Some producers of microcontrollers such as Parallax are making components for specific use with Java.

Python

Python is one of the most popular languages for scripting. It is easy to learn and could be used to quickly and efficiently integrate several programs.

If you have selected a hobbyist type of microcontroller from a known producer, there's a chance that you can find a book that you can read so you can learn how to program in their preferred programming language. But if you instead prefer a microcontroller from a smaller producer, it is crucial to see what language the controller wants to use and what tools are available.

Conclusion

Thank you again for purchasing this book!

I hope this book was able to help you to learn the basic building blocks of robot building.

The next step is to expand your knowledge in robotics, especially learning advanced programming for your robot.

Finally, if you enjoyed this book, please take the time to share your thoughts and post a review on Amazon. It'd be greatly appreciated!

Thank you and good luck!

Book 2

Human-Computer Interaction

By Solis Tech

The Fundamentals Made Easy!

Human-Computer Interaction: The Fundamentals Made Easy!

Table of Contents

Introduction

I want to thank you and congratulate you for purchasing the book, "Human-Computer Interaction: The Fundamentals Made Easy!"

This book contains proven steps and strategies on how to conceptualize and design a computer system that incorporate principles on effective interaction between the user and the device.

Human-computer interaction (HCI) is the study of the interaction between people and computers and the degree at which computers are developed enough to successfully interact with humans. So many institutions particularly academic and corporations now study HCI. Unfortunately, ease-of-use has not been a priority to most computer systems developer. The issue continues to bedevil the HCI community as accusations still abound that computer makers are still indifferent and are not making enough effort to make truly user-friendly products.

On the other hand, the designing task of computer system developers is not simple either as computers are very complex products. It is also true that the demand for use of computers have grown by leaps and bounds outstripping the need for ease-of-use by a significant margin. If you are a computer designer or simply have basic interest in making devices more effective for users, this book will help you a lot.

Thanks again for purchasing this book, I hope you enjoy it!

Chapter 1: Aspects of HCI

Main aspects of HCI

HCI is composed of three main features, namely: the user, the computer, and their interaction or how they work together.

"User" refers to either the individual or the group of users doing things together. An understanding of how the people's sense of sight, hearing, and touch send information is very important. Also, the type of mental models of interactions differs according to the personality of the user. And finally, interactions are also influenced by cultural and national differences.

"Computer," on the other hand, pertains to all technology from desktop to huge computer systems. As an example, if the topic is website design, the computer would then be the website. "Computers" would also include gadgets like mobile phones or even VCRs.

Finally, the "Interaction" is what happens as "User" uses the "Computer" to achieve a certain objective. Humans, of course, are totally different from machines. So the HCI's main intent is to ensure a successful interaction between the two.

In this aspect, adequate knowledge about humans and computers are critical to realize a functioning system. You need to seek inputs from users. Such knowledge would provide much needed information in determining schedule and budget that are crucial to the systems. In effect there are ideal situations and perfect systems. But the key is finding the balance between what is ideal and what is really feasible given the existing situation.

Objectives of HCI

HCI aims to come up with systems that are functional, usable, and safe. Developing computer systems with excellent usability depends on:

- having enough understanding of the aspects that lead people to use technology in certain ways
- being able to devise tools and ways for creating suitable systems
- the development of safe, effective, and efficient interaction

- making people the priority

The main philosophy underneath HCI is that the users or the people using the computers always come first. Developers must always be guided by the users' needs and preferences in designing systems. It is the system that should match the requirements of human users and not people changing to suit the nature of the machines.

The primacy of usability

Usability is one of the principal considerations in HCI. It is simply about ensuring that a system can be easily learned and used or be what is called user-friendly. A system is considered usable if it:

- can be learned easily
- can be remembered easily in terms of use
- is effective
- is efficient
- is safe
- is enjoyable

Lack of usability means wasted time, mistakes, and disappointments. Unfortunately, a lot of existing systems and devices have been designed without sufficient attention to usability. These include ATM, the Web, computer, printer, mobile phone, personal organizer, coffee machine, remote control, soft drink machine, ticket machine, photocopier, stereo, watch, video game, library information systems, and calculator.

A good example is the photocopier. If you are not familiar with the symbols on the buttons you will be greatly confused. For instance, the big button with the C on it actually refers to Clear, not Copy. The button used to produce copies is actually on the left side with an unrelated symbol. Devices and gadgets should be easy, effortless, and enjoyable to use.

Analyzing and designing a system based on HCI principles involve a lot of factors that produce really complex analysis because of interactions among many of them. The major factors are:

- The User – motivation, satisfaction, experience, enjoyment, personality. Also cognitive processes and capabilities

- <u>User Interface</u> – navigation, output devices, icons, commands, input devices, graphics, dialogue structures, user support, use of color, multimedia, natural language
- <u>Environmental Factors</u> – health and safety, noise, heating, lighting, ventilation
- <u>Organization Factors</u> – job design, work organization, training, roles, politics
- <u>Task Factors</u> – task allocation, skills, easy, novel, complex, monitoring
- <u>Comfort Factors</u> – seating, layout, equipment
- <u>Constraints</u> – budgets, buildings, cost, equipment, timescales, staff
- <u>Productivity Factors</u> – decrease costs, increase quality, increase innovation, increase output, decrease errors
- <u>System Functionality</u> – software, hardware, application

There are different disciplines representing a wide array of subjects that are covered in HCI. The manifold inputs from these fields have continued to enrich HCI. The disciplines include:

- Cognitive Psychology – limitations, performance predictions, information processing, cooperative working, capabilities
- Ergonomics – display readability, hardware design
- Computer Science – graphics, software design, prototyping tools, technology, User Interface Management Systems (UIMS) and User Interface Development Environments (UIDE)
- Social Psychology – social and organizational structures
- Engineering and Design – engineering principles, graphic designs
- Linguistics – natural language interfaces
- Philosophy, Sociology, and Anthropology – computer supported cooperative work (CSCW)
- Artificial Intelligence – intelligent software

Chapter 2: The Human Side in the HCI

Some of the key aspects that shed light on the human side of HCI are:

1. Perceptual-Motor Interaction. Effective human-computer interface design requires an appreciation of the whole human perceptual-motor system. The information-processing approach is central to the perceptual-motor behavior study and for considering the human factors in HCI. An effective interface design reflects the designer's knowledge of the perceptual such as visual displays, use of sound, and graphics. Also the cognitive exemplified by conceptual models and desktop metaphors as well as motoric constraints like ergonomic keyboards of the human perpetual-motor system.

   Man has gone beyond the use of computer punch cards and command-line interfaces. We now use speech recognition, eye-gaze control, and graphical user interfaces. The importance of various perceptual, cognitive, and motor constraints of the human system is now better recognized in HCI. An effective interface must take into account the perceptual and action expectations of users, the action that is seen with a response location, and the mapping of the perceptual-motor workspaces.

2. Human Information Processing. Aspects of human information processing such as models, theories, and methods are currently well developed. The available knowhow in this field is broadly useful to HCI in general such as in the representation and communication of knowledge and visual display design. An effective HCI requires making the interaction compatible with the human information-processing capabilities. Many things about human information processing have been integrated into cognitive architectures that are now applicable to HCI. These applications include the Model Human Processor, the Act model, the SOAR model, and the Epic model.

3. Mental Models in Human–Computer Interaction. Studying mental models can help understand HCI by inspecting the processes by which such models impact behavior. For example, mental models of machines can enable both novice and seasoned problem solvers to find new methods for fulfilling a task through more elaborate encoding of remembered methods.

   The Reverse Polish Notation is a great example of this. There is also a general theory that says readers develop a representation in their mind at several levels of what they read. First is the encoding of text, followed by the representation of propositional content of text. Finally, to this text, they integrate world knowledge to form a mental model of the situation described.

Readers also have the ability to look for ideas in multiple texts. They construct a kind of structured mental maps that show which documents contained which ideas even when they did not expect to need it while reading. Mental models are generally considered as semantic knowledge. Focusing on the degree of commonality among team members, for instance, when it comes to knowledge and beliefs, allows quantitative measures of similarity and differences which is the language of computers.

4. Emotion in Human–Computer Interaction. Emotion used to be persona non grata in the field of computer design. It had no place in the efficiency and rationality of computers which were the personification of zero emotion. Recent study findings in the field of psychology and technology show in a totally different light the relationship between humans, computers, and emotions.

Emotion has ceased to be considered only in light of anger generated by inexplicable computer crashes or hyper excitement caused by video games. Nowadays, it is widely accepted that a host of emotions are important part of computer-related activities such as Web search, sending an email, online shopping, and playing computer games. In almost everything now, the emotional systems get engaged according to psychologists.

Studies and discussions on emotion and computers have grown a lot because of dramatic advances in technology. Computers have actually been used to evaluate the relationship between emotion and its correlates. In the same vein, the astounding improvements in quality and speed of signal processing now enable computers to form conclusion on a user's emotional condition. Compared to purely textual interfaces that have very limited range, the multimodal interfaces that can use voices, faces, and bodies are now more capable to a broader range of emotions.

Nowadays, the performance of an interface will be seriously impeded without considering the user's emotional state. Surprisingly, it can earn even descriptions like socially inept, incompetent, and cold. Much remains to be done to successfully incorporate emotion recognition into interfaces. Still, more studies about the interaction between design and testing can help create interfaces that are efficient and effective while providing satisfaction and enjoyment.

5. Cognitive Architecture. A cognitive architecture is a computer simulation program that makes use of human cognition principle based on human experimental data. It also refers to software artifacts developed by computer programmers. Likewise, the term also includes large software systems which are considered hard to develop and maintain.

Right now, cognitive architectures are not widely utilized by HCI practitioners. Nevertheless, it is quite relevant as an engineering field to usability and has important applications in computing systems especially in HCI. It also serves as theoretical science in human computer interaction studies. Finally, cognitive architectures combine artificial intelligence methods and knowledge with data and principles from cognitive psychology.

Presently-known cognitive architectures are undergoing improvements and are being utilized in HCI-related tasks. Two of the most well-known systems, EPIC and ACT, are production systems or built around one. All systems have production rules which differ from architecture to architecture. The difference lies on focus and history although there's a certain similarity in intellectual history. They may have more congruence than differences at some levels either because of mutual borrowings or due to the convergence of the science. The third system, Soar, is a bit different than the first two production system models.

The three production systems, Soar, Epic, and ACT-R were developed to present different types of human cognition but showed more similarities than divergence as they developed. It is not easy to describe a value possessed by architecture as advantage because to others it constitutes a disadvantage. For instance, Soar's learning mechanism is very important for modeling the improvement of users for a period of time. But there are many applications also where Soar's features result to harmful side effects that can cause more difficulty in model construction.

6. Task Loading and Stress in HCI. Stress in the form of task loading is central to HCI. The traditional perspective on stress sees it in light of exposure to some adverse environmental situations such as noise and the focus of attention centers on most affected physiological system. A new way of looking at it, however, stems from the findings that all stress effects are mediated through the brain.

And since the brain is mainly focused on ongoing behavior or current task, stress ceases to be a peripheral issue but that the ongoing task becomes the primary source of stress. And this renders stress concerns that are central to all HCI issues. This means computer-based systems which aim at helping people lessen cognitive workload and task complexity actually impose more burdens and stress on them.

The person's coping mechanism for such stress affects their work performance and personal wellbeing. The environment may vary but some mechanisms for appraising stress in all task demands are the same. So for HCI, certain principles and designs for stress are applicable across multiple domains.

40

There are several theories of stress and performance and their connection to human-computer interaction. Workload and stress are at times considered as varying perspectives on the same problem. There are some general practices for stress mitigation. But quite important for this topic is setting up effective measures of information processing and mental resources. It also includes expounding on task dimensions that are relevant and their relationship to self-regulatory mechanisms.

It is critical to establish how an individual's appraisal of his/her environment can be influenced by personal traits and states. This is because stress can only be understood vis-a-vis interaction between a person and the environment. Lastly, it is better to treat stress at multiple levels whether physiological or organizational when making practical application. Instead of one-dimensional which is bound to fail, multidimensional is better as it considers the person, task, and the physical, social and organizational environments.

The implication is that HCI researchers and practitioners should go beyond the design of interface displays and controls and focus also on the person aspects. What are the things in the individual that affect performance and the physical-social environment where the human-technology interaction happens? It means that the technical principles at work in that situation are not adequate. They cannot develop a complete description of the relationship between resources and cognitive activities.

7. Motivating, Influencing, and Persuading Users. From its former role as tool for scientists, the spread of computer use to all sectors of society has brought new uses for computers. Among those uses are persuading people to change their attitudes and behavior. Nowadays, it is widely accepted that skills in motivating and persuading people are necessary for developing a successful HCI.

Interaction designers are actually agents of influence which unfortunately they have not yet understood and applied. Yet their works often involve creating something that tries to change people though they may not be conscious of it. Among these works are motivating people to register the software, learn an online application, or have product loyalty. Changing people's attitudes is now a common feature in the success of interactive products.

Depending on the types of product, the persuasion factor can either be small or large. At any rate, anything that needs to be marketed needs to be persuasive. The growing use of computing products and the limitless scalability of software makes interaction designers one of the best potential change agents in the future.

Take for example the Web-interaction designers who increasingly are facing more challenge in designing something that will hold the attention and motivation of information seekers. After that, they need to persuade web users to adopt certain behaviors like:

- using a software
- joining a survey
- clicking on the ads
- returning often after bookmarking a site
- buying things online
- releasing personal information
- forming an online community

Being able to persuade people is a measure of success here. But with success comes responsibility. The Web designer needs to make the website credible. The following are some broad guidelines to ensure credibility:

1. Design websites to present the real and practical aspects of the organization.
2. Invest sufficiently in visual design
3. Make websites that people can easily use.
4. Include markers of good quality
5. Use markers of reliability.
6. Avoid too much commercialism on a website
7. Adopt and adjust to the user experience
8. Avoid being amateurish

To sum it all, computer systems have become an inescapable part of everyday life. The interactive experience involving all systems be it mobile phone or desktop can be designed in such a way as to influence the way we think and act. By combining the computing capability with persuasion psychology, computer systems can motivate and persuade. Humans are undoubtedly still superior when it comes to influencing people. But in many areas of endeavor, computer can do what humans cannot even imagine being capable of.

Computers don't sleep and can be designed to keep trying on and on. At the very least, computers provide a new way for modifying how people act and think. Like it or not, the community of HCI professionals is at the forefront of the campaign to make more sensitive and responsive tech products. It can rise to the challenge of helping churn out products that enhance the people's over-all quality of life. Or it can continue being a tool to produce mindless products whose main reason for being into is to make profit for the owners.

8. Human-Error Identification in Human–Computer Interaction. The leap from focusing on human error in technological problems to a less obvious culprit started in the 1940's. It was established during that year that plane pilot error was often designer error. It began to show that design is the key to

42

substantially reduce human error and this paradigm continued to gather steam particularly in HCI.

It is now common wisdom that human error can be as often as the product of a defective design or as a person making a mistake. The inadequate design fosters activities that lead to errors. A groundbreaking outcome of this new philosophy is that errors are now viewed as totally predictable events instead of seeing it as unpredictable occurrences. This makes errors avoidable.

So errors became instances where planned series of steps and activities fail to realize intended results independent of any outside change agencies. If errors are no longer random, then it can be identified and predicted ahead of time. What partially drove this line of thinking are the accidents that happened in the nuclear industry that is hungry for preemptive solutions. This has led to the formulation of several human-error identification (HEI) techniques.

Although evaluative and summative in nature, these HEI techniques that employ ergonomics methods can now be used in formative design stages especially in analytic prototyping. For instance, the entry of computer-aided design such as in architecture has profound impacts on prototyping. It made possible what was considered as impossible or too prohibitively costly design alteration at the structural prototyping stage.

The three main forms of prototyping human interfaces have been identified namely: functional analysis, scenario analysis, and structural analysis.

Functional analysis includes consideration of the functional range supported by the device. In comparison, scenario analysis is exemplified by consideration of the device in relation to events sequence. An example of the structural analysis, on the other hand, is the use of user-centered viewpoint in a non-destructive testing of the interface.

One compelling example of the crucial role of design in predicting and minimizing errors concern human error identification (HEI) tools like the TAFEI or Task Analysis for Error Identification. The results of the application of TAFEI on interface project designs show how it can improve systems and its relevance to other ergonomic methods. It served to validate what has been long suspected when it comes to error-design relationship as follow:

- Structured systems like TAFEI results to reliable and trustworthy error data;

- Most errors resulting from technology are totally predictable;

- To improve design and reduce errors, ergonomics methods should be employed in formative design process.

Exploring design weaknesses through tools like TAFEI will go a long way in developing and producing devices and gadgets that are tolerant to error.

Chapter 3: The Computer Side in the HCI

The salient points when it comes to the computer side in HCI include:

1. <u>Input Technologies and Techniques</u>. Input devices which are also a classification of computer can detect physical aspects of places, things, and, of course, people. However, its function is never complete without considering the visual feedback corresponding to the input. It is like using a writing instrument without something to write on. Input and output should always go together.

   And in devices with small screens, this is only possible with the help of integrated sensors. If the user or human characteristics are important in a maximized HCI environment, so are input technologies with enough sophistication to meet user-machine interaction requirements. Users can only achieve the task objectives by combining the right feedback with inputs. In this regard, the HCI designer should take into account the following:

a. the industrial and ergonomic design of the gadget

b. the physical censor

c. the relationship among all interaction techniques

Input gadgets have many properties that apply to the usual pointing devices or mobile items with touch input. These pointing devices include the: mice, trackballs, isometric joysticks, isotonic joysticks, indirect tablets, touchpads, touchscreens, and pen-operated devices. The mice or mouse, of course, is one of the most popular as anyone who has ever used a computer knows. Because of its inherent advantages for individual users where it can easily be used by most people, it is one of the most preferred pointing devices.

Touchpads are most well-known to laptop users. These are small tablets that are sensitive to touch and which are usually featured on laptop units. Touchscreens on the other hand are tablets that are sensitive to touch which are placed on a display. It is increasingly becoming the tool to beat because of the proliferation of smart phones and other hand-held devices.

There are input models and theories that are quite helpful in evaluating the efficacy of interaction strategies. But it would be most beneficial to readers here to focus on current and future trends for this feature. Interactive system designers should go beyond the usual things like graphical user interface and pointing ideas when it comes to inputs.

They must delve deeper into more effective search strategies, sensor inputs for new data types, and techniques of synthesis to make much better sense of data. Better search tools will enhance navigation and manual search regarding file systems. One outstanding development is the breakthrough in the development of more advanced sensor inputs such as technologies for tagging and location.

It allows computers to identify physical objects and locations that have been tagged, and to detect their location and distance to other devices through signal strength analysis. These sensors are making interface personalization much easier. This development in interaction also has great implications for data mining and techniques for machine learning. Continuous improvement in structure synthesis and extraction techniques is invaluable in this data-rich era.

An overriding aim in HCI is to achieve dramatic advancement in humanity's interaction with technology. The computer side of this presents limitless possibilities but the cognitive skills and senses of man will be relatively stagnant. Our holding, touching, and object-movement are not the result of technology-like progress but a product of our human limitations.

2. Recognition- and Sensor-Based Input for Interaction. Computers are able to manipulate physical signals that have been transformed by sensors into electrical signals. Sensors have found their way into various fields of industry such as robotics, automotive, and aerospace. It has also found vast applications in consumer products.

    The computer mouse is a very good example. Imagine that simple-looking device equipped with algorithms that process images and specialized camera that enables it to be unbelievably sensitive to motions. It detects movement at the rate of a thousand of an inch several thousand times per second. Another interesting device is the accelerometer that detects acceleration due to movement and continuous acceleration because of gravity.

    Digital cameras now make use of accelerometers to save a photo. Laptops also are equipped with accelerometers for self-protection. When the laptop is accidentally dropped the accelerometer enables the hard disk to secure the hard drive prior to impact. With smartphones, the goal is for motion sensing for the purpose of interaction such as determining the walking pattern of users. Generally, HCI research on sensors dwells on its usage to improve interaction.

    Sensor studies are either to broaden input options or build new computing forms. The new forms include mobile devices that recognize locations and places that are sensitive to the presence and needs of its inhabitants. Still there are far more advanced goals and applications like in robotics.

There is a race to develop machines that will behave and think like humans or at least complement their capabilities. It has many critical applications such as in nuclear power accident mitigation. One worry, of course, is that it will end up in the military. But in safety, mobile computing, entertainment, productivity, affective computing, and surveillance, sensors are finding widespread application.

An intriguing side note here is the idea of developing a sensor to enable computers to detect and accordingly react to the frustration of its user. The computer's response could be something like playing relaxing music. Sensing could be in the form of the user banging on the keyboard in frustration. A microphone could react to the yelling of the speaker or the webcam could sense scowling.

In general, the potential of interactive sensing is quite good. The degree of progress across the whole computing spectrum actually gives the impression that sky is the limit. Advances in nanotechnology, CPU power, and storage capacities will continue to produce more outstanding innovations in the computer side of HCI.

But what is driving the unprecedented growth of the sensor-based interactive systems is the dizzying expansion in devices outside the old desktop computer. It is hard to keep track of the explosive proliferation of smart phones, tablet PCs, portable gaming devices, music and movie players, living room-centric computers, and personal digital assistants. Computing is becoming part and parcel of our daily life and our environment.

Through recognition techniques and sensing systems, task-specific computing devices will be developed instead of general functions. It will also pave the way for different types of interaction style in HCI. This activity-specific interactive systems development will further hasten innovations on a much broader array of practical applications.

3. Visual Displays. Timekeeping has always been one function that man has strived for a good visual display. Today's smartwatches which are actually wrist computers sport stunning visual displays. It is no longer limited to displaying time but is multifunctional. Some brands can pinpoint exact location in the planet through a global positioning system. Others can show heart rate while a number can be personal digital assistants.

The main idea behind wearable computing is that the human body is wearing the visual displays. One major way people use wearables is to put the display on one's head making user's hands free to work. It is called headmounted displays or HMDs. The screen-based is one category of HMDs. It makes use of the retinal-projection method which projects images on the retina of the eye.

An alternative method is the scanning displays which scan images onto the retina pixel by pixel. A second type of display which is actually much bigger in scope and the most widespread is the hand-held and wrist-worn displays. They are in mobile phones, media players, wristwatches, and other portable gadgets. Apple is one of the global leaders in this field and the most well-known. Even textiles for clothing are now being used for such technology in what is now known as photonic textiles-fabric.

Multicolored lighting systems were merged with the fabrics for its electronic information function without affecting the cloth's softness. It has sensors, GSM, and Bluetooth! Photonic fabrics have great promising applications in the areas of personal health care and communication.

4. <u>Haptic Interfaces</u>. Haptic interface refers to a device for sending feedback that produces sensation of weight, touch, rigidity, and other aspects through the skin and muscle. This force feedback mechanism is designed to enhance computer-human interaction. Because haptics are done through actual physical contact, they are not easy to synthesize unlike the sense of sight and sound that are gathered through the eyes and ears.

   The genius in the haptic interface is that it simply makes use of the body's own highly sophisticated receptor system. The haptic feedback is made possible through synthetic stimulation in the skin and proprioception.

   Proprioception involves something deeper than the skin – the muscle and skeleton. The mechanoreceptors in the body enable its detection of contact forces received from the environment. Body receptors sense velocity, skin stretching, vibration, and edges of objects. Haptic interfaces are more widely applied in the field of virtual reality than in information media and related devices are now available commercially.

Two of the most important research needs on haptic interfaces in the future concern the psychology in haptics and safety considerations. Safety is a crucial issue as insufficient actuator control can lead to injuries for users. Control problems may occur with the tool displays and exoskeleton. Unintended forces or vibration may pose danger to the user.

A locomotion interface that holds a user's body can cause serious physical damage if control is inadequate. It requires proven safety equipment that amply protects the walker and this should be a major objective of research. A much safer alternative is a system where the user does not wear any equipment during the interaction.

The psychology in haptics on the other hand requires more studies on muscle sensation as most existing findings are on skin sensation. Among the few promising findings relate to Laderman and Klatzky's work (1987) on force display and their recent study of forces distributed according to space. Their psychological findings have very promising applications in the development of haptic interface. A lot of obstacles need to be overcome before usage of haptic interface becomes widespread.

Though men cannot do without haptics in real life interaction, it is still of limited use in HCI. One can say haptic interface is still in infancy with its 10-year background. So eventually its time will come just like image displays (e.g. TV and movies) which started 100 years ago. For now, what are available are a few haptic interfaces with limited functionality and high cost. At the very least, haptic interface is a new very promising frontier in HCI with immense potential contribution to man's quality of life.

5. Non-speech Auditory Output. Sound is one of the key aspects that complete our interaction with our environment. But where speech is direct and necessitates focus and attention, non-speech sound is more diffused and provides a different class of information.

   Non-speech sounds include sounds from the environment, music, and sound effects. Nevertheless, speech and non-speech sounds complement each other just like text is complemented by visual symbols. Non-speech sounds can give information in a shorter period of time than speech.

   Right now, the non-speech field needs more research. The user interface is a much more effective tool for HCI when it employs a combined visual and sound feedback. This sound-visual combination has complementary function as well. Visual gives specific information about a small area reached by our eyes but sound or the auditory system provides more general information from beyond our focus.

   Our senses are the key to our effective interaction with the external world. These senses in turn bring more dimensions in information as they enhance one another. These principles are very useful in a multimodal HCI by adding non-speech sound output to the graphical displays. An example of this application is focusing our eyes on one task like editing a manuscript while monitoring other aspects in the machine through sound.

Reliance on visual sense which is more prevalent at present can be problematic. One problem is there could be visual overload which means the user could miss lots of information. Or simply that the viewer cannot look at everything at the same time at all times. Sound can help eliminate that situation by giving information to the user that the eyes could not see.

This interdependence between visual and audio could make information presentation far more efficient. Non-speech sound is mainly used in games' sound effects, music, and other multimedia usages. It is commonly employed in creating a certain mood for the item like in movies. In HCI, sound is used to provide information particularly those things that a user does not see or notice such as what is going on in their computer systems.

It is useful to use non-speech sound in HCI for many reasons. Seeing and hearing in the human body is first of all interdependent. The eyes can give information that is high-resolution only in a limited area of focus. But sounds can be received from all sides of the user: front, above, below, and behind. This not only provides direct information but also tells the eyes where to look to get more useful data. In fact, at times reaction to sound stimuli is faster than what is seen.

Non-speech sound can therefore help in reducing large display overload which can cause users to miss important data. This is especially true in large graphical interfaces that use multiple monitors. Using sound to present some information would reduce screen space. It would also lessen the volume of information that should be on the screen. This is most relevant to gadgets with small visual displays like smartphones and PDAs.

Non-speech sound would also decrease demands on our visual attention. For instance, a user who is walking would miss much information as he looks at his device's visual display because of competing attention from the traffic or uneven surface where he is walking. In fact, if the information is in sound, he does not have to look at his device at all.

Our sense of sound is also underutilized. Yet as exemplified by classical music, its intricate organization can make, say a symphony, a powerful tool for transmitting complex information. The beauty with sound is that it grabs attention. It is easy to avoid looking at something but hard to ignore sound which makes it very effective in sending important information. Likewise, certain things in the interface look more natural in sound than in sight.

Finally, non-speech sound will allow visually-impaired users to use computers. Newer graphical displays have, in particular, made it even harder for them to operate the device. Research has been extensive in the HCI application of non-speech sound in a wide range of topics.

There are two main areas of growth where the application of non-speech sound has the best potential. One is in the creation of multimodal displays that utilizes all available senses. This means integrating sound with other things like force-feedback and tactile apart from sight. The other area is in wearable and mobile computing gadgets that also use multimodal displays.

As mentioned, the screens of these devices are small and sound will reduce the need for screen space.

6. <u>Network-Based Interaction</u>. Networked interfaces have modified our perception of society and the world at large particularly with the Web and now mobile devices. There are several roles that networks play in HCI. The first is as an Enabler which refers to things that can be done only with network. The second is as Mediator which pertains to problems and issues caused by networks.

Third is as Subject which focuses on managing and understanding networks and fourth as Platform which dwells on interface architectures and algorithms. Network includes both the wire-based and the wireless world. Things are rapidly changing especially in the wireless networks. These changes can be classified in two dimensions, namely:

- **Global vs. Local** – refers to the distance by space between the connected points such as machines in the office to global networks like the Internet.

- **Fixed vs. Flexible** – pertains to the nature of the links between points such as fixed devices and gadget that configures itself. More changes are coming because of spreading wireless links. One example is being able to gain access to internet connections and printers of another office by simply plugging a portable device into the Ethernet.

Traditionally, LANs belong to local-fixed category while Internet is global-fixed. Hand-helds like cell phones are also categorized as global-fixed because phones are fixed and independent of location. The internet makes use of domain names which are fixed like URLs. Some phone technologies like GSM and GPRS are classified as global-fixed because it is possible to send content that is based on location. Also the enlarging data capability is enabling services to handle huge media content.

What set these technologies apart, however, are the connectivity model and the charging which are usually by data use or fixed charge. There are a number of current and new technologies from the local-flexible type. These include the Wi-Fi, infrared, Bluetooth, and ZigBee which permit flexible connections among personal gadgets. With them, a computer device can utilize a mobile-phone modem or a headset with Bluetooth can make connections with a phone, wireless. Unfortunately, these capabilities also enhance unsavory activities like illegal equipment accessing, hacking, and surveillance.

7. <u>Wearable Computers</u>. Computers have become like appendage to many office workers. But it is hard for those using mobile devices to get the information

they need. In a mobile situation, existing interfaces will hamper the user's main task. Users will be forced to prioritize the device instead of the environment. The need is for a wearables design that helps fulfill not obstruct the task.

A framework that can be very useful in creating good designs of computer interfaces which are wearable is CAMP. This framework addresses different factors that may impinge on the effectiveness of the design such as body closeness and how it is used. CAMP stands for:

- <u>Corporal</u> – which means absence of discomfort to users during physical interface with the wearable.
- <u>Attention</u> – interface design should allow user to focus both on the real world and virtual reality.
- <u>Manipulation</u> – there are adequate controls which are easy to manipulate particularly in a mobile environment.
- <u>Perception</u> – Design must enable user to quickly perceive displays even when mobile. So displays should be easy to navigate and simple.

Outside offices and buildings, an attractive option for a user to have access to a computer interface is through wearables. There are challenges however that need to be addressed to fulfill the tasks in terms of contextual awareness, interface, adaptation to tasks, and cognitive model. These include:

- **Modalities of Input/output** – the ease of use and accuracy of modalities developed that try to copy the human brain's input/output capacity are not yet satisfactory. Frustrations bedevil users when there are inaccuracies. Also the computing requirement of these modalities is way beyond what low-weight wearable devices have. Input devices which are simple to use are needed.

- **Models of User interface** – there is a need for extensive experimentation in using applications involving end-users.

- **Capability-applications matching** – evaluation and design of interface should prioritize development of most effective way to access information and avoid creating additional features.

- **Simple methodology in interface evaluation** – current evaluation approaches are too complicated and time-consuming making them unsuitable in interface design. What is needed is an evaluation methodology that addresses frustration and human errors.

- **Context awareness** – for context aware computing to be realized, several questions must be answered. These include application models that integrate the social and cognitive aspects, social and cognitive mapping of inputs from many sensors, anticipating the needs of users, and interacting with the users.

Conclusion

Thank you again for purchasing this book!

I hope this book was able to help you to gain useful knowledge and understanding about human-computer interaction.

The next step is to apply what you have learned.

Finally, if you enjoyed this book, please take the time to share your thoughts and post a review on Amazon. It'd be greatly appreciated!

Thank you and good luck!

Book 3
Quality Assurance
By Solis Tech

Software Quality Assurance Made Easy!

Quality Assurance: Software Quality Assurance Made Easy!

Table Of Contents

Introduction

I want to thank you and congratulate you for purchasing the book, *"Quality Assurance: Software Quality Assurance Made Easy"*.

This book contains proven steps and strategies on how to implement Software Quality Assurance.

Software quality assurance evolved from the quality assurance in industrial production in the 1940s. With the introduction of computers and the development of applications, it has become an important aspect of software development. In this book, you will learn about the various concepts of software quality assurance and its application in your workplace.

Chapter 1 talks about the definitions of software quality assurance and quality control. It is necessary that you understand what it is so that you can fully grasp the other fundamental concepts. Chapter 2 discusses the ways you can implement software quality assurance in an existing environment. In Chapter 3, you will learn how you can ensure that the software produces quality outputs.

Chapter 4 teaches you how to develop and run software testing while Chapter 5 talks about the timing of the software release. In Chapter 6, you will learn about using automated testing tools.

Thanks again for purchasing this book, I hope you enjoy it!

Chapter 1: **Definition Of Software Quality Assurance And Software Testing**

Software quality assurance is part of the process of software development. It includes the improvement and monitoring of the process. It also follows the processes, procedures, and standards. Furthermore, it ensures the resolution of problems discovered. In effect, it focuses on the prevention of problems.

Software testing, on the other hand, includes the operation of the application or system under controlled conditions, which must include all abnormal and normal conditions. The primary goal of software testing is to know what can go wrong when unexpected and expected scenarios occur. Its focus is on the detection of problems.

Companies differ in assigning responsibilities for quality assurance and software testing. There are organizations that combine both responsibilities to a single person or group. However, it is also common to have various teams that combine developers and testers. Project managers lead such teams. The decision to either combine or separate responsibilities depends on the company's business structure and size. Human beings or machines can perform software testing.

Some projects do not require a group of testers. The need for software testing depends on the project's context and size, the development methodology, the risks, the developers' experience and skill, and other factors. For example, a short-term, low-risk, and small project with expert developers, who know how to use test-first development or thorough unit testing, may not require software test engineers.

In addition, a small or new IT company may not have a dedicated software testing staff even its projects have a need for it. It may outsource software testing or use contractors. It may also adjust its approach on project development and management by using either agile test-first development or more experienced programmers. An inexperienced project manager may use his developers to do their own functional testing or skip thorough testing all together. This decision is highly risky and may cause the project to fail.

If the project is a huge one with non-trivial risks, software testing is important. The use of highly specialized software testers enhances the ability of the project to succeed because different people with different perspectives can offer their invaluable contribution because they have stronger and deeper skills.

In general, a software developer focuses on the technical issues to make a functionality work. On the other hand, a software test engineer focuses on what can go wrong with the functionality. It is very rare to come across a highly effective technical person, who can do both programming and testing tasks. As such, IT companies need software test engineers.

There are various reasons why software may have bugs. First, there is no communication or miscommunication as to the application's requirements. Second, the software is very complex so an inexperienced individual may have a difficult time understanding the software application. In addition, the software may be a big one and includes big databases, data communications, security complexities, different remote and local web services, and multi-tier distributed systems.

Third, there are programming errors. Fourth, the requirements of the software change because the end-user requests for them. In some cases, this end-user may not be aware of the effects of such changes, which can result in errors. The staff's enthusiasm is affected. Continuous modification requirements are a reality. Quality assurance and software test engineers must be able to plan and adapt continuous testing to prevent bugs from affecting the software.

Fifth, time pressures can be difficult to manage. Mistakes often occur during crunch times. Sixth, people involved in the project may not be able to manage their own egos. Seventh, the code has poor documentation or design. It is difficult to modify and maintain codes that are poorly written or documented. Bugs can occur if IT companies do not offer incentives for developers to write maintainable, understandable, and clear codes. Most companies offer incentives when they are able to finish the code quickly. Eighth, the use of software development tools can also result to bugs.

How Quality Assurance Evolved

In 1946, the US Occupation Forces established the Quality Movement in Japan. The movement follows the research of W. Edwards Deming and the Statistical Process Control papers. The methods discussed by the research and papers pertained to industrial production. Each process of production has an output with a required specification and a verifiable main product. The Quality Control group measures the output at different production stages. It ensures that the output falls within the acceptable variances.

The Quality Assurance group does not part in the production process. It audits the process to ensure compliance with the established standards and guidelines. It gives its input for the continuous improvement of the process. In a manufacturing setup, it is easy to differentiate between quality assurance and quality control. These methods become the norm in manufacturing. Since they work in industrial production, they spawned the birth of software quality control and software quality assurance.

Quality Attributes Of Software

Hewlett Packard's Robert Grady developed the common definition of the Software Quality Attributes. The FURPS model identified the following characteristics: functionality, usability, reliability, performance, and supportability (FURPS).

The functional attributes pertain to the features of the software. They answer the question about the purpose of the software instead of its use. The reason for the software's existence is different from the concerns about its reliability, look and feel, and security. The usability attributes are characteristics pertaining to user interface issues like consistency, interface aesthetics, and accessibility.

The reliability attributes include recoverability, accuracy, and availability of the system while performance attributes pertain to issues like startup time, recovery time, system response time, and information throughput. Supportability addresses issues like localizability, scalability, installation, configurability, compatibility, maintainability, adaptability, and testability. The FURPS model evolved into FURPS+ to include specification of constraints like physical, interface, implementation, and design constraints.

The Software Quality Control team tests the quality characteristics. The tests for usability and functionality occur during execution of the actual software. On the other hand, adaptability and supportability tests occur through code inspection. It is not the responsibility of the Software Quality Control or Software Quality Assurance to put the quality attributes into the software. The Software Quality Control team tests for the absence or presence of such characteristics while the Software Quality Assurance group ensures that each stakeholder follow the right standards and procedures during software execution.

In theory, the implementation of FURPS+ will overcome the problems caused by the software's intangible nature because the Software Quality Control team can measure each software attribute. For example, the amount of time it takes for programmers to fix a bug is a measure of supportability. To improve it requires the implementation of new coding standards.

The Software Quality Control group can inspect the code to ensure compliance with the coding standard while the Software Quality Assurance team can ensure that the quality control and programmer teams follow the right standards and process. It is the duty of the Software Quality Assurance group to collect and analyze the time spent on fixing the bug so that it can provide an input in terms of its usefulness to the process improvement initiative.

Chapter 2: Introducing Software Quality Assurance Procedures To An Existing Company

The introduction of the software quality assurance procedures rely on the risks involved and the company's size. If the organization is large and the projects are high-risk, the management must consider a more formal quality assurance process. On the other hand, if the risk is lower, the implementation of quality assurance may be a step-at-a-time procedure. The processes of quality assurance maintain balance with productivity.

If the project or the company is small, an ad-hoc process may be more appropriate. The success of the software development relies on the team leaders and developers. More importantly, there must be enough communications between software testers, managers, customers, and developers. Requirement specifications must be complete, testable, and clear. There must be procedures for design and code reviews, as well as retrospectives. Common approaches used are Agile, Kaizen, and Deming-Shewhart Plan-Do-Check-Act methods.

It is important to evaluate documents, code, specifications, plans, and requirements. To do this, it is necessary to prepare issues lists, checklists, inspection meetings, and walkthroughs. This process is what IT people refer to as verification. On the other hand, validation occurs after verification and includes actual testing of the application. A walkthrough is evaluation in an informal meeting. There is no need to prepare for it.

An inspection is more formal than a walkthrough and attended by at most eight people including a reader, a recorder, and a moderator. Usually, an inspection's subject is a test plan or requirements specification so that the attendees will know the software's problems. The attendees must read the document before attending an inspection so they will find the problems prior to the meeting. It is difficult to prepare for an inspection but it is an effective and least costly way to ensure quality.

Testing Requirements That Must Be Considered

The basis of black box testing is the software's functionality and requirements. A software tester need not be knowledgeable of code or internal design. White box testing, on the other hand, requires knowledge of code. It takes into consideration the code statements, paths, branches, and conditions. Unit testing requires testing of particular code modules or functions. Usually, the application developer performs it because it requires detailed knowledge of the code and program design.

API testing is testing of data exchange or messaging among systems parts. An incremental integration testing tests the application when there is a new functionality, which must be independent to work separately before it can be included in the application. The application programmer or software tester performs this type of testing. An integration testing requires the testing of the various parts of the software in order to know if they work together properly. Functional testing, on the other hand, is a kind of black box testing that focuses on the application's functional requirements. The software testers perform it.

System testing is also another type of black box testing. It includes testing of the general requirements specifications and all the various parts of the software. End-to-end testing, on the other hand, is a macro type of system testing. It includes testing the whole software in various situations that replicates real-world use. Smoke testing or sanity testing is an initial testing to know if the new version is performing well so that major testing can commence. Regression testing retests the application after some modifications or fixes.

Acceptance testing is final testing the software based on customer's specifications. Load testing is software testing under heavy loads in order to find out when the response time of the system fails or degrades. Stress testing, on the other hand, is system functional testing with unusually heavy loads, large complex queries of the database, using large numerical values, and heavy repetition of certain inputs or actions. Performance testing is testing using test or quality assurance plans.

The goal of usability testing depends on the customer or end-user. It uses techniques like surveys, user interviews, and user sessions' video recording. Software testers and developers are not the people to implement usability testing. Install/uninstall testing uses processes to test upgrade, full, or partial install/uninstall of the software. Recovery testing determines how the software can recover in catastrophic instances like hardware failures and crashes. Failover testing is another name for recovery testing.

Security testing tests how the application protects against willful damage, unauthorized access, and the likes. It uses sophisticated testing methods. Compatibility testing tests how the application performs in some environments. Exploratory testing, on the other hand, is an informal and creative testing that does not use any formal test cases or plans. Usually, software testers are new to the application. Ad-hoc testing is almost the same as exploratory testing but the software testers understand the application prior to testing.

Context-driven testing focuses on the software's intended use, culture, and environment. For example, medical equipment software has a different testing

63

approach than a computer game. User acceptance testing determines if the application is satisfactory to the customer or end-user. Comparison testing, on the other hand, compares the software with its competitors.

Alpha testing occurs when the software is almost finished. Usually, end users perform this type of testing. Beta testing is testing when the software is finished and ready for final release. Like alpha testing, this kind of testing is for end users. Finally, mutation testing uses test cases and allows for code changes and retesting using the original test cases to discover software bugs.

Common Problems And Solutions In Developing Software

Software development can encounter common problems like poor user stories or requirements, unrealistic schedule, inadequate testing, featuritis, and miscommunication. Problems can arise if there requirements are incomplete, unclear, not testable, or too general. Furthermore, if the software development has a short timetable, problems can also be evident. If the software does not pass through the testing process, the IT organization will only know of problems if the system crashes or the end users complain.

In addition, problems can arise if the users or customers request for more new functionalities even after the development goals are set. Finally, if there is miscommunication between the programmers and customers, problems are inevitable. If it is an agile project, the problems become evident when it strays away from agile principles.

If there are common problem then there are also common solutions like solid requirements, realistic schedules, adequate testing, sticking to original requirements, and communication. Software requirements must be complete, clear, cohesive, detailed, testable, and attainable. All players concerned must agree upon them. In agile environments, there must be close coordination with end users or customers so that any change in requirement is clear. There must be enough time for design, planning, bug fixing, testing, changes, retesting, and documentation. Each person must not feel burnout during the project.

For testing to be adequate, it must start early. Retesting must occur after each change or fix. There must be enough time for bug fixing and testing. If the end user or customer requests for excessive changes when development has begun, it is important that the project manager explain the consequences. If the change is important, a schedule change is inevitable to accommodate it.

The project manager must manage the expectations of the end users or customer. In an agile environment, the project manager can expect significant change in the initial requirements. There must be a schedule for inspections and walkthroughs. The use of communication tools is important to ensure cooperation and teamwork.

Software Quality Assurance And Software Quality Control In The Workplace

If there is no formal implementation of Software Quality Assurance/Software Quality Control, the quality group can mirror that of an engineering services group. It performs the tasks not done by the development team. It reports to the application development manager. Under this scenario, the manager can instruct the quality assurance personnel to install a load-testing tool and declare him as a performance expert. Although the quality assurance person must perform load testing, it is the job of the software designer/programmer. As such, the software quality assurance/software quality control must be separate from the development team.

Software quality assurance requires the establishment of templates used during reviews. Such templates must have sections for both non-functional and functional requirements. For example, the requirements for performance must be in terms of transaction rates and user population. The use of Traceability Matrix can help in the requirements management. The matrix also encourages the analyst to use individual requirements for cross-referencing.

Software quality assurance verifies that the requirements comply with the templates and that they are not ambiguous. It reviews the risks of non-completion of the non-functional attributes, as well as the Traceability Matrix, to ensure usability of all requirements in other specifications. Software quality assurance can write the test cases, which refer to some of the requirements. It can cross-reference the requirements in the Traceability Matrix.

The interface specification is a requirement if the software is component-based. Software quality assurance must subject the document to verification. It must also subject other lower level specifications to quality assurance if such requirements are critical. Even if there are limited resources for software quality assurance, an IT organization can obtain a good return-on-investment if it pays attention to requirements and interfaces.

Chapter 3: **Ensuring Software Quality**

Software quality is the delivery of bug-free software within budget and on time. Because quality is subjective, it depends on the customer and his overall influence in the project. A customer can be an end user, a customer contract officer, a customer acceptance tester, a customer manager, a software maintenance engineer, and others. Each customer has his own definition of quality.

A code is good if it works, is secure, bug free, maintainable, and readable. Some companies have standards, which their programmers follow. However, each developer has his own idea about what is best. Standards have different metrics and theories. If there are excessive rules and standards, they can limit creativity and productivity. A design can be functional or internal design.

An internal design is good if the overall structure of the software code is clear, easily modifiable, understandable, and maintainable. It is robust with the capability to log status and with sufficient error handling. Furthermore, the software works as expected. A functional design is good if the software functionality comes from the requirements of the end user and customer. Even an inexperienced user can use the software without reading the user manual or seeking online help.

The software life cycle starts with the conception of the application and commences when the software is no longer used. It includes factors like requirements analysis, initial concept, internal design, functional design, test planning, documentation planning, document preparation, coding, maintenance, testing, integration, retesting, updates, and phase out.

A good software test engineer must possess a test to break attitude, which means that he must be able to focus on the detail in order to ensure that the application is of good quality. He must be diplomatic and tactful so that he can maintain a good relationship with the programmers, the customers, and the managements.

It is helpful if he has previous experience on software development so that he understands the whole process deeply. A good software test engineer is able to appreciate the point of view of the programmers. He must have good judgment skills to be able to determine critical or high-risk application areas that require testing.

A good quality assurance engineer also shares the same qualities with a good software test engineer. Furthermore, he must understand the whole process of software development and the organizations' goals and business approach. He must possess understanding of all sides of issues in order to communicate effectively with everyone concerned. He must be diplomatic and patient, especially in the early stages of the quality assurance process. Lastly, he must be able to detect problems during inspections and reviews.

Qualities Of A Good Quality Assurance Or Test Manager

A good manager must be familiar with the process of software development and must be enthusiastic and promote positivity in the team. He must be able to increase productivity by promoting teamwork and cooperation between quality assurance, software, and test engineers. If the processes require improvements, he must be diplomatic in ensuring smooth implementation of such changes.

He must be able to withstand pressure and offer the right feedback about quality issues to the other managers. He must be able to hire and keep the right personnel. He must possess strong communication skills to deal with everyone concerned in the process. Lastly, he must stay focused and be able to hold meetings.

Importance Of Documentation In Quality Assurance

If the team is large, it is more useful to have the proper documentation for efficient communication and management of projects. Documentation of quality assurance practices is necessary for repeatability. Designs, specifications, configurations, business rules, test plans, code changes, bug reports, test cases, and user manuals must have proper documentation. A system of documentation is necessary for easy obtaining and finding of information.

Each project has its own requirements. If it is an agile project, the requirements may change and evolve. Thus, there is no need for detailed documentation. However, it is useful to document user stories. Documentation on the software requirements describes the functionality and properties of the software. It must be clear, reasonably detailed, complete, attainable, cohesive, and testable. It may be difficult to organize and determine the details but there are tools and methods that are available.

It is important to exercise care in documenting the requirements of the project's customer, who may be an outside or in-house personnel, an end user, a customer contract officer, a customer acceptance tester, a customer manager, a sales person, or a software maintenance engineer. If the project did not meet the expectations, any of these customers can slow down the project.

Each organization has its own way of handling requirements. If the project is an agile project, the requirements are in the user stories. On the other hand, for other projects, the requirements are in the document. Some organizations may use functional specification and high-level project plans. In every requirement, documentation is important to help software testers to make test plans and perform testing. Without any significant documentation, it is impossible to determine if the application meets the user expectations.

Even if the requirements are not testable, it is still important to test the software. The test results must be oriented towards providing information about the risk levels and the status of the application. It is significant to have the correct testing approach to ensure success of the project. In an agile project, various approaches can use methods that require close cooperation and interaction among stakeholders.

Chapter 4: How To Develop And Run Software Testing

To develop the steps for software testing, it is necessary to consider the following.

First, it is important to get the requirements, user stories, internal design, functional design, and any other information that can help with the testing process.

Second, testing must have budget and schedule.

Third, it must have personnel with clear responsibilities.

Fourth, there must be project context to determine testing approaches, scope, and methods.

Fifth, identification of limitations and scope of tests is necessary in order to set priorities.

Sixth, testing must include methods and approaches applicable.

Seventh, it must have the requirements to determine the test environment.

Eighth, it must have requirements for testing tools.

Ninth, it is important to determine the data requirements for testing input.

Tenth, identification of labor requirements and tasks are important.

Eleventh, it must determine milestones, timelines, and schedule estimates.

Twelfth, testing must include error classes, boundary value analysis, and input equivalence classes, when needed.

Thirteenth, it must have a test plan and documents for approvals and reviews.

Fourteenth, testing must determine test scenarios and cases.

Fifteenth, it must include reviews, approvals of test cases, inspections, approaches, and scenarios.

Sixteenth, there must be testing tools and test environment, user manuals, configuration guides, reference documents, installation guides, test tracking processes, archiving and logging processes, and test input data.

Seventeenth, it must include software releases,

The Test Plan

A test plan is a document, which includes scope, objectives, and focus of the software testing activity. To prepare it, it is important to consider the efforts needed to validate the software's acceptability. The test plan must help even those people who are not part of the test group. Testing must be thorough but not very detailed.

The test plan can include the following: the title, the software version number, the plan's revision history, table of contents, the intended audience, the goal of testing, the overview of the software, relevant documents, legal or standards requirements, identifier and naming standards, and requirements for traceability. The test plan must also include overall project organization, test organization, dependencies and assumptions, risk analysis of the project, focus and priorities of testing, limitations and scope, test outline, data input outline, test environment, analysis of test environment validity, configuration and setup of test environment, and processes for software migration among other things.

A test case includes the input, event, or action plus its expected result in order to know if software's functionality is working as planned. It contains test case identifier, objective, test case name, test setup, requirements for input data, steps, and expected results. The details may differ, depending on the project context and organization. Some organizations handle test cases differently. Most of them use less-detailed test scenarios for simplicity and adaptability of test documentation. It is important to develop test cases because it is possible to detect problems and errors in the software through them.

When A Bug Is Discovered

If the software has a bug, the programmers must fix it. The software must undergo retesting after the resolution of the error. Regression testing must be able to check that the solutions performed were not able to create more problems within the application. A system for problem tracking must be set up so it becomes easier to perform the retesting. There are various software tools available to help the quality assurance team.

In tracking the problems, it is good to consider the following: complete information about the bug, bug identifier, present bug status, software version number, how the bug was discovered, specifics about environment and hardware, test case information, bug description, cause of the bug, fix description, retesting results, etc. There must be a reporting process so that the appropriate personnel will know about the errors and fixes.

70

Quality Assurance: Software Quality Assurance Made Easy!

A configuration management includes the various processes used to coordinate, control, and track requirements, code, problems, documentation, designs, change requests, tools, changes made, and person who made the changes. If the software has many bugs, the software tester must report them and focus on critical bugs. Because this problem can affect the schedule, the software tester must inform the managers and send documentation in order to substantiate the problem.

The decision to stop testing is difficult to do, especially for complex applications. In some cases, complete testing is not possible. Usually, the decision to stop testing considers the deadlines, the degree of test cases completion, the depletion of the test budget, the specified coverage of code, the reduction of the bug rate, and the ending of alpha or beta testing periods. If there is not enough time to perform thorough testing, it is important to focus the testing on important matters based on risk analysis. If the project is small, the testing can depend on risk analysis again.

In some cases, there are organizations that are not serious about quality assurance. For them, it is important to solve problems rather than prevent problems. Problems regarding software quality may not be evident. Furthermore, some organizations reward people who fix problems instead of incentivizing prevention of problems. Risk management includes actions that prevent things from happening. It is a shared responsibility among stakeholders. However, there must be a point person who is primarily responsible for it. In most cases, the responsibility falls on the quality assurance personnel.

Chapter 5: Deciding When To Release The Software To Users

In most projects, the decision to release the software depends on the timing of the end testing. However, for most applications, it is difficult to specify the release criteria without using subjectivity and assumptions. If the basis of the release criteria is the result of a specific test, there is an assumption that this test has addressed all the necessary software risks.

For many projects, this is impossible to do without incurring huge expenses. Therefore, the decision is actually a leap of faith. Furthermore, most projects try to balance cost, timeliness, and quality. As such, testing cannot address the balance of such factors when there is a need to decide on the release date.

Usually, the quality assurance or test manager decides when to release the software. However, this decision involves various assumptions. First, the test manager understands the considerations, which are significant in determining the quality of the software to justify release. Second, the quality of the software may not balance with cost and timeliness. In most organization, there is not enough definition for sufficient quality. Thus, it becomes very subjective and may vary from day to day or project to project.

The release criteria must consider the sales goals, deadlines, market considerations, legal requirements, quality norms of the business segment, programming and technical considerations, expectations of end users, internal budget, impact on the other projects, and a host of other factors. Usually, the project manager must know all these factors.

Because of the various considerations, it may not be possible for the quality assurance manager or test manager to decide the release of the software. However, he may be responsible in providing inputs to the project manager, who makes the release decision. If the project or the organization is small, the decision to release the software rests on the project manager or the product manager. For larger organizations or projects, there must be a committee to decide when to release the software.

If the software requirements change continuously, it is important for all stakeholders to cooperate from the beginning so that they all understand how the change in requirements may affect the project. They may decide on alternate strategies and test plans in advance. It is also beneficial to the project if the initial design of the software can accommodate changes later on. A well-documented code makes it easier for programmers to make the necessary changes.

Quality Assurance: Software Quality Assurance Made Easy!

It is also good to use rapid prototyping if possible so that customers will not make more requests for changes because they are sure of their requirements from the very beginning. The initial schedule of the project must allow extra lead-time for these changes. If possible, all changes must be in the "Phase 2" of the software version. It is important that only easily implemented requirements must be in the project.

The difficult ones must be in the future versions. The management and the customers must know about costs, inherent risks, and scheduling impacts of big changes in the requirements. If they know about the challenges, they can decide whether to continue with the changes or not.

There must be balance between expected efforts with the effort to set up automated testing for the changes. The design of the automated test scripts must be flexible. Its focus must be on aspects of application, which will remain unchanged. The reduction of the appropriate effort for the risk analysis of the changes can be through regression testing.

The design of the test cases must be flexible. The focus must be on ad-hoc testing and not on detailed test cases and test plans. If there is a continuous request for changes in the requirements, it is unreasonable to expect that requirements will remain stable and pre-determined. Thus, it may be appropriate to use approaches for agile development.

It is difficult to determine if software has significant hidden or unexpected functionality. It also indicates that the software development process has deeper problems. The functionality must be removed if does not serve any purpose to the software because it may have unknown dependencies or impacts. If not removed, there must be a consideration for the determination of regression testing needs and risks. The management must know if there are significant risks caused by the unexpected functionality.

The implementation of quality assurance processes is slow over time because the stakeholders must have a consensus. There must be an alignment of processes and organizational goals. As the organization matures, there will be an improvement on productivity. Problem prevention will be the focus. There will be a reduction of burnout and panics. There will be less wasted effort and more improved focus.

The processes must be efficient and simple to prevent any "Process Police" mentality. Furthermore, quality assurance processes promote automated reporting and tracking thereby minimizing paperwork. However, in the short run, implementation may be slow. There will be more days for inspections, reviews, and planning but less time for handling angry end users and late night fixing of errors.

If the growth of the IT organization is very rapid, fixed quality assurance processes may be impossible. Thus, it is important to hire experienced people. The management must prioritize issues about quality and maintain a good relationship with the customers. Every employee must know what quality means to the end user.

Chapter 6: Using Automated Testing Tools

If the project is small, learning and implementing the automated testing tools may not be worth it. However, these automated testing tools are important for large projects. These tools use a standard coding language like Java, ruby, python, or other proprietary scripting language. In some cases, it is necessary to record the first tests for the generation of the test scripts. Automated testing is challenging if there are continuous changes to the software because a change in test code is necessary every time there is a new requirement. In addition, the analysis and interpretation of results can be difficult.

Data driven or keyword driven testing is common in functional testing. The maintenance of actions and data is easy through a spreadsheet. The test drivers read the information for them to perform the required tests. This strategy provides more control, documentation, development, and maintenance of test cases. Automated tools can include code analyzers, coverage analyzers, memory analyzers, load/performance test tools, web test tools, and other tools.

A test engineer, who does manual testing, determines if the application performs as expected. He must be able to judge the expected outcome. However, in an automated test, the computer judges the test outcome. A mechanism must be present for the computer to compare automatically between the actual and expected outcomes for every test scenario. If the test engineer is new to automated testing, it is important that he undergo training first.

Proper planning and analysis are important in selecting and using an automated tool for testing. The right tool must be able to test more thoroughly than manual methods. It must test faster and allow continuous integration. It must also reduce the tedious manual testing. The automation of testing is costly. However, it may be able to provide savings in the long term.

Choosing The Appropriate Test Environment

There is always a tradeoff between cost and test environment. The ultimate scenario is to have a collection of test environments, which mirror exactly all possible usage, data, network, software, and hardware characteristics that the software can use. For most software, it is impossible to predict all environment variations. Furthermore, for complex and large systems, it is extremely expensive to duplicate a live environment.

The reality is that decisions on the software environment characteristics are important. Furthermore, the choice of test environments takes into consideration logistical, budget, and time constraints. People with the most technical experience and knowledge, as well as with the deep understanding of constraints and risks, make these decisions. If the project is low risk or small, it is common to take the informal approach. However, for high risk or large projects, it is important to take a more formalize procedure with many personnel.

It is also possible to coordinate internal testing with efforts for beta testing. In addition, it is possible to create built-in automated tests upon software installation by users. The tests are able to report information through the internet about problems and application environment encountered. Lastly, it is possible to use virtual environments.

The Best Approach To Test Estimation

It is not easy to find the best approach because it depends on the IT organization, the project, as well as the experience of the people involved. If there are two projects with the same size and complexity, the right test effort for life-critical medical software may be very large compared to a project involving an inexpensive computer game. The choice of a test estimation approach based on complexity and size may be applicable to a project but not to the other one.

The implicit risk context approach caters to a quality assurance manager or project manager, who uses risk context with past personal experiences to allocate resources for testing. The risk context assumes that each project is similar to the others. It is an experience-based intuitive guess.

The metrics-based approach, on the other hand, uses past experiences on different projects. If there is already data from a number of projects, the information is beneficial in future test planning. For each new project, the basis of adjustment of required test time is the available metrics. In essence, this is not easy to do because it is judgment based on historical experience.

The test-work breakdown approach decomposes the testing tasks into smaller tasks to estimate testing with a reasonable accuracy. It assumes that the breakdown of testing tasks is predictable and accurate and that it is feasible to estimate the testing effort. If the project is large, this is not feasible because there is a need to extend testing time if there are many bugs. In addition to, there is a need to extend development time.

The iterative approach is for large test projects. An initial rough estimate is necessary prior to testing. Once testing begins and after finishing a small percentage of testing work, there is a need to update the testing estimate because testers have already obtained additional knowledge of the issues, risks, and software quality. There is also a need to update test schedules and plans. After finishing a larger percentage of testing work, the testing estimate requires another update. The cycle continues until testing ends.

The percentage-of-development approach requires a quick way of testing estimation. If the project has 1,000 hours of estimated programming effort, the

IT firm usually assigns a 40% ratio for testing. Therefore, it will allot 400 hours for testing. The usefulness of this approach depends on risk, software type, personnel, and complexity level.

In an agile software development approach, the test estimate is unnecessary if the project uses pure test-driven development although it is common to mix some automated unit tests with either automated or manual functional testing. By the nature of agile-based projects, they are not dependent on testing efforts but on the construction of the software.

Conclusion

Thank you again for purchasing this book!

I hope this book was able to help you to understand the concepts of Software Quality Assurance.

The next step is to apply what you learned from this book to your work.

Finally, if you enjoyed this book, please take the time to share your thoughts and post a review on Amazon. It'd be greatly appreciated!

Thank you and good luck!

www.ingramcontent.com/pod-product-compliance
Lightning Source LLC
Chambersburg PA
CBHW060412190526
45169CB00002B/873